The Effect of Land Use on Biodiversity

A Stormwater Pond and Zooplankton Study

By: Michanna Talley, JD, MS, BS, Cert.

Published by Jazi Gifts by Michanna, LLC, Publishing Division

Greenville, South Carolina

www.jazigifts.com

1-888-778-4808.

ISBN 978-0-9842684-0-5

The Effect of Land Use on Biodiversity

A Stormwater Pond & Zooplankton Study

TABLE OF CONTENTS

Prologue

In our current society, there is a great push to be "eco-friendly." We are constantly provided information regarding the human effect on the Earth. However, there is little public information which looks separately at the effect of commercial and residential areas on the environment when compared to open space areas (those with little to no human intrusion). Although several environmental peer reviewed journals do exist, the public is not generally privy to this information. Additionally, the articles in those journals rarely cater to the public and therefore are generally not understood by the public as they are not in layman's terms. This book takes an immensely scientific study and provides the same information to the everyday person.

Chapter 1:

Why is this study important?

Changes have been forced onto our environment as a consequence of urbanization. One such change, increased surface runoff, is due to actions such as construction of residential and commercial areas. Following this type of development, flooding is much more likely. The water can no longer infiltrate into the soil due to paved roads and constructed buildings. Therefore the water ends up flooding areas in which it finally settles. In order to combat this flooding, stormwater ponds are constructed. Their primary water source is simply runoff following precipitation. This precipitation may encounter either a detention or retention pond (commonly observed simply as lakes within residential neighborhoods, apartment complexes, and business parks).

Detention ponds were designed to only store water for a short period of time and to reduce the rate of the follow of water back into the ground. Retention ponds, also referred to as extended detention ponds or wet basins, store water indefinitely. This allows an improvement in water quality due to sedimentation by which solid debris as well as pollutants settle out of the water[14, 4].

Biological uptake refers to the removal of dissolved metals and nutrients by the various organisms present in constructed ponds. This is usually aided by aquatic plants which impede the flow of the water into the pond and trap pollutants. These plants actually increase biological uptake[3]. The biological and chemical processes in stormwater ponds may actually mirror those of natural lakes due to the permanent pool of water and sediments found in retention ponds.

Natural wetlands provide flood control and improve water quality. They are usually densely populated with several types of vegetation[2]. Many retention ponds are relatively young compared to natural wetlands, which convey a long term record of past environmental conditions within the bottom sediment. Water enters a wetland environment by at least one of three sources, direct precipitation, surface water runoff, and groundwater[2]. Surface water and precipitation both contribute to the standing water of stormwater retention ponds while groundwater does not usually account for any water volume.

Within aquatic systems, many types of organisms are found. Zooplankton are one common type of organism. They can be thought of as aquatic insects. Zooplankton can be generally broken down into four groups: a) protists, b) rotifers, c) cladocera, and d) copepods. However, "true" zooplankton only include rotifers, cladocera, and copepods. Zooplankton can also be broken down into macrozooplankton and microzooplankton. Macrozooplankton include copepods, the genus *Chaoborus*, and cladocera. Microzooplankton include rotifers. Microzooplankton are usually smaller than one millimeter and macrozooplankton are greater than one millimeter.

Compared to other types of zooplankton, protists have a small biomass. Protists are further broken down into flagellates and ciliates. Protists feed on bacteria, algae, and other protists and some are able to carry out photosynthesis. Rotifers include categories of both nonpredatory and predatory organisms. They may feed on bacteria, small algae, protozoa, or other rotifers. Both cladocera and copepods are carnivorous or herbivorous. Their reproductive rates are based on both temperature and food availability. Predation mostly increases in the summer while their own population changes with the presence of invertebrates and fish. Their mortality rate may also correlate with their migratory responses[15]. In order to avoid predation, zooplankton remain in dark areas or among vegetation. This shelters them from fish which feed on zooplankton.

Currently, little is known about zooplankton in stormwater ponds. Their biodiversity within these ponds may be due to several factors such as the presence or absence of fish, vegetative cover (plants covering a section of the water), riparian cover (trees and shrubs on the bank of the body of water), and the size of the pond. Additionally, the ecosystem type in which these ponds are found may have an effect. In general, zooplankton biomass is not solely controlled by fish predation unless predation intensity is very high. There is however a positive correlation between phytoplankton and zooplankton.

Pond biodiversity is greater than that of rivers, lakes, and streams, although the richness (the amount of organisms) is variable among ponds. In addition, many ponds do not have many species in common and therefore have several unique species[17]. Water body size is an obvious characteristic affecting diversity. Surprisingly, studies have found that several small ponds have a larger total biodiversity than that of one larger body of equal total area[9]. Other known factors that have an effect on diversity include riparian cover , vegetative cover, the current season, and the presence or absence of fish.

During early spring, copepods have been found to be most abundant when samples are taken from the deepest part of lakes[10]. During the summer, evaporation may lead to the exclusion of some species. Those ponds with or without aquatic vegetative cover usually have varying species numbers and dominance due to the strong relationship between diversity and percentage of pond open surface area, an area usually in the center not covered by vegetation[18]. The aspect known to affect zooplankton communities the most is the presence or absence of fish. There is a distinct difference found in the organisms present in ponds with and without fish. The best predictor of diversity in many studies is the presence of stocked fish[19]. Simply put, an increase in fish number is known to decrease zooplankton food availability. Conversely, food availability has an effect on the growth rate of the fish that are present. Piscivorous fish (fish that feed on other fish) decrease the number of zooplanktivorous fish (fish that feed on zooplankton) therefore increasing zooplankton, which in turn decreases algae. Although this is generally true, it is not always the case. A decrease of rainbow trout has been shown to actually be compensated for by other fish as well as copepods, which may become even more abundant than before. This change may also lead to an increase of algae[5].

From studies of natural lakes and ponds, expectations related to zooplankton (density, diversity, and spatial structure) can be made for stormwater ponds. However, one must keep in mind that stormwater ponds are subject to different types of runoff due to the landscape environment in which the stormwater ponds are found (ex: residential, commercial, or open space).

Human disturbance is also known to affect zooplankton communities. In a study conducted by Yakovlev in 2001, zooplankton were studied in two polluted bodies of water, Monche Bay, containing metals (nickel, copper, cadmium, cobalt, zinc, and chromium), suspended particles (sulfate and chlorine), toxic chemicals, as well as oil products, and Belaya Bay, containing sulfate, chlorine, fluorine, aluminum, iron, strontium, and titanium[20]. Within both ecosystems, *Bosmina* (Cladocera) and *Asplanchna priodonta* (Rotifera) were found. Additionally in Belaya Bay, *Notholca caudate*, a rotifer, was found. Qualitatively, increased densities of "poor" or tolerant zooplankton were found near polluted sites and rotifer and cladoceran diversity and abundance increased away from these discharge sites. In the discharge areas, the tolerant zooplankton made up 90% of the total biomass with the number of sensitive species being very low (*L. kindtii, B. cederstromii*, and *Heterocope*). Aside from chemical pollution, predaceous copepods are tolerant of high water turbidity (cloudiness of the water) and oxygen depletion.

A previous study by Karouna-Renier and Sparling (2001), found that water chemistry remained relatively constant among 20 stormwater ponds, divided based on adjacent land use[8]. Temperature however did vary, with residential pond temperature being the lowest and commercial pond temperature lower than highway pond temperature. The major aim of that study, measurement of metal geochemistry in the ponds (i.e. copper, zinc, calcium, & magnesium), did not differ among the land use types in the aqueous form nor in the sediment. Odonates (includes dragonflies and damselflies) from commercial ponds were found to have higher copper and zinc levels, correlating with aqueous levels. The level of copper in odonates was related to land use and sampling period, as was lead in mollusks (organism typically with shells). The overall levels of copper and zinc in commercial ponds were two times that of the open space ponds. Larger invertebrates were found to have low levels of lead which was positively correlated with turbidity, caused by increased sedimentation and urbanization. During wet seasons, an increased pollutant loading was found to have occurred leading to higher metal concentrations.

Still, little or nothing has been published on zooplankton communities in constructed wetlands, especially in developed areas. These ponds are a relatively new entity, however with time, their numbers will increase, possibly outnumbering natural wetlands. These ponds no longer simply retain water, but have become their own ecosystem.

Chapter 2:

How was this study conducted?

The objectives of this research were to 1) describe the diversity and abundance of zooplankton communities in stormwater ponds and 2) examine the relationship between adjacent land use and zooplankton community structure.

A "natural" experiment was carried out to determine the effect that land use changes have on the zooplankton community of stormwater ponds in residential, commercial, and open space areas. All the ponds except for one, came from an existing database of 20 stormwater retention ponds in Prince George's, Anne Arundel, and Howard Counties, located in Maryland that have been the topic of intense study of amphibian and benthic invertebrate communities (communities on the water bottom)[8]. An additional residential pond was added to the study due to a minimum water level needed, not available at other residential ponds, as well as sampling availability. For all ponds, the presence or absence of fish, which can have an effect on zooplankton communities, was also known[12,13].

At each pond site, pond characteristics, such as visible human use, vegetative cover, and riparian cover values were recorded. For both riparian and vegetative cover, each pond was visibly split into four quadrants. Each quadrant was then scored (on a scale of one to one hundred) in terms of the amount of riparian cover on the bank and the amount of the pond covered with vegetation. Weather was also observed and recorded. Once on the pond (in the open water zone), pond depth as well as secchi depth was recorded. For secchi depth (a measure of the distance into the water that sunlight can reach), a secchi disk (a circular tool shaded black and white) was lowered into the pond. The depth at which the disk could no longer be seen was recorded. The disk was then slowly brought up to the surface of the water. The depth at which it could just again be seen was also recorded. The average of these two depths was recorded as the secchi depth. Readings of temperature, conductivity (the presence or absence of dissolved solids), and dissolved oxygen were recorded with the use of a hydrolab multi-probe, an instrument specially created for this purpose. This probe was introduced into the water and allowed to adjust to new conditions before readings were recorded.

A total of seven samples were taken from each pond. Three zooplankton samples were obtained with the use of a plankton trap which allowed for the analysis of zooplankton from a set amount of water (30 liters in this study). Another sample, obtained with the use of a zooplankton tow net (a net towed behind a boat), was obtained to capture more zooplankton taxa. Three additional samples were taken for further water chemistry analysis, chlorophyll analysis, and pH readings. Once off the pond, all samples were preserved for further experimentation. Once in the lab, samples were frozen until ready for analysis.

From the water chemistry samples from each pond, both ammonium and phosphorus concentrations were determined. Samples were allowed to thaw and reach room temperature prior to analysis. Samples with known ammonium concentrations were used alongside pond samples to determine accurate concentration values. As with ammonia determination, samples of known phosphorus concentration were also used. In order to estimate the amount of algae in each pond, chlorophyll analysis was completed. Following analysis, chlorophyll concentrations were calculated.

Both macrozooplankton and microzooplankton were analyzed from each zooplankton sample and identified. All samples were diluted to equal volumes. From each diluted sample, ½ milliliter of each sample was viewed with a microscope. Zooplankton were then counted and identified with the use of the Center for Freshwater Biology Research Zooplankton Key[6]. From the final zooplankton values, densities from the samples were calculated. Once all zooplankton were identified, counted, and recorded, three types of diversity were determined. These types are as follows: 1) alpha diversity, the number of organisms within a particular ecosystem, 2) beta diversity, the total number of unique organisms found when ponds were compared to one another, and 3) gamma diversity, the overall total number of organisms found[16].

From all obtained zooplankton values, one way ANOVAs (an analysis of variance where it is determined whether or not significant differences exist) were completed to determine differences between land uses, if any. Multivariate analyses (a technique used to model and analyze several variables) were also done.

Chapter 3:

What information was gained?

Several of the ponds seemed to have no visible human use. Others were used to dump unwanted items. One pond, Fairland, an open space pond, has been made into a scenic area. Several people were walking around the pond. One pond in particular, Briar Oak, is very active. The community treats it as a recreational fishing hole. Several boats were on shore, hinting to the fact that it is used for recreation.

In terms of depth and secchi depth, the open space ponds, Fairland and Bowie Fleming had both the largest average depth and secchi depth. However, a commercial pond, Laurel Lakes Court, had a secchi depth identical to that of Bowie Fleming. Overall, residential ponds had the lowest secchi depth.

Riparian cover (found on the banks of the ponds) was rather full for all ponds. Vegetative cover however was not. Two commercial ponds, Laurel Employment Park and Golden Triangle had no vegetative cover. The largest average vegetative cover was found at Country Meadows, a residential pond.

Conductivity values were found to fall within the same large range for all ponds. The lowest value was recorded at a commercial pond, with the highest value found at a residential pond. The open space pond values were the most constant. Both the commercial and residential group had one pond with a very low dissolved oxygen level.

Temperature values found did not have a large range (22.3 – 29.1 °C). On average, the commercial ponds had the highest temperature. Other ponds, such as Muirkirk Road (residential) and Bowie Fleming (open space) also had a relatively high temperature. All pH values were around 7.0 (neutral). The lowest pH values belonged to residential ponds and the highest pH value to a commercial pond.

One pond, Golden Triangle, a commercial pond had no chlorophyll. The largest chlorophyll value was assigned to a residential pond, Muirkirk Road. Phosphorus values were relatively highest for Laurel Employment Park, a commercial pond. The highest ammonium concentration belonged to a commercial pond, Golden Triangle.

In terms of samples at open space ponds, Fairland had only copepods. Bowie Fleming, also an open space pond, was found to have rotifers, cladocera, and copepods. Overall, copepods were most dense, followed by rotifers and cladocera. For commercial ponds, Golden Triangle had the largest density of cladocera, followed by equal densities of rotifers and copepods. The next commercial pond, Laurel Lakes Court, was the only pond found to have *Chaoborus* (which do not fit into any of the major zooplankton categories). The *Chaoborus* were the greatest in density, followed by copepods and then rotifers. No cladocera were found in this pond. The last commercial pond, Laurel Employment Park, had a very large copepod density value, followed by rotifers and then cladocera. Among residential ponds, Country Meadows was found to have only rotifers and copepods, with copepods having the greatest density. Briar Oak, the next residential pond, had mostly rotifers, followed by cladocera and then copepods. Muirkirk, the remaining residential pond, had a very large density of rotifers, then copepods, and finally cladocera.

In terms of overall density per liter, Muirkirk, a residential pond, was found to have the greatest value and Country Meadows, also a residential pond, was found to have the lowest value. Commercial ponds also had varying results. Most all open space samples were dominated by copepods.

When looking only at average zooplankton abundance based solely on land use, in both open space and commercial ponds, copepods were in greatest abundance, followed by rotifers, and then cladocera. In residential ponds, rotifers had the greatest abundance, followed by cladocera, copepods, and then *Chaoborus*.

When zooplankton were categorized into macrozooplankton and microzooplankton, new information arose. Overall, all residential ponds were dominated by microzooplankton. Two of the commercial ponds (Laurel Lakes Court and Laurel Employment Park) were also dominated by microzooplankton. Golden Triangle (commercial) however, was dominated by macrozooplankton. In the open space pond samples, one pond, Bowie Fleming, was dominated slightly by microzooplankton, while the other, Fairland, had an equal density of macrozooplankton and microzooplankton.

When all data from each pond was compiled and then grouped based on land use, open space ponds were found to have the lowest alpha diversity, followed by commercial, and then residential ponds, with the most. When commercial ponds were compared to residential, there were 14 unique groups of organisms between the two (beta diversity). In the comparison between residential and open space ponds and between commercial and open space ponds, both had 13 unique species between the two. From these values, differences between each land use seem relatively consistent. From the entire study of all three land uses, 29 groups of organisms were found (gamma diversity).

Diversity values were also calculated within each land use. Alpha diversity for commercial ponds ranged from four to 11. A total of 18 types of zooplankton were found between the three commercial ponds. Overall, 23 types of zooplankton were found in the residential ponds. Between the two open space ponds, a total of 12 types of zooplankton were found, four in one and 10 in the other. When compared to each other, these open space ponds had one of the lowest beta diversity values when compared to others in the study.

Interactions between zooplankton abundance (density) and chemical variables were analyzed. At times there were weak relationships and sometimes no relationship. Overall average density when compared with dissolved oxygen values showed that the greatest densities were found in ponds with the highest recorded oxygen content.

The density of each major group of zooplankton was also compared to chemical values. When analyzed alongside secchi depth, average rotifer density was greatest in ponds with a lower depth through which light is able to penetrate. Average rotifer density, similarly to overall average density, was found to be highest in ponds with a larger dissolved oxygen value.

Copepod density was found to coordinate with several variables. Riparian cover, although not a chemical variable, was found to have an interaction with average copepod density. An optimum riparian cover of about 75% produced larger copepod numbers. Copepod densities were also higher when lower conductivity values were recorded. Although varied, densities of copepods were greater when oxygen content values were higher. Chlorophyll levels were also related to copepod densities. When phosphorus concentration was compared to copepod density, a very straightforward relationship emerged (greater concentrations of phosphorus paralleled greater copepod density).

Chaoborus, which was only found in one pond, also showed a peculiar interaction with pH. It was only found in the pond with the highest pH value. Conversely, *Chaoborus* density was greatest in the pond with the lowest recorded conductivity. Diversity of each pond was also compared with chemical variables. At greater oxygen levels, greater diversity was found. Also, lower conductivities produced the greatest diversity.

ANOVAs were performed to determine any difference in land uses with regards to diversity (alpha diversity) as well as abundance (density). Abundance of each zooplankton type was also analyzed. The only significance found related to rotifer density and land use. Residential ponds had significantly higher rotifer density then either open space or commercial ponds. No difference in rotifer density was found between open space and commercial ponds.

Multivariate analyses were used to compare environmental data, as well as the presence or absence of genera. The greatest similarity was present among open space ponds with the greatest difference between the residential ponds. With environmental data coupled with the presence or absence of the various types of zooplankton, open space ponds were found to be most similar with residential ponds once again showing the greatest difference.

Chapter 4:

Where is the data?

Table 1: Year of Construction & Known Animal Presence; NA – not available[8, 12]

Pond	Land-use	Year	Presence
Laurel Employment Park	Commercial	1991	Fish & Amphibian
Golden Triangle	Commercial	1991	NA
Laurel Lakes Court	Commercial	1990	Amphibian
Briar Oak	Residential	1985	Fish & Amphibian
Muirkirk Road	Residential	1990	Fish & Amphibian
Country Meadows	Residential	1989	Fish
Fairland	Open Space	1989	Fowl
Bowie Fleming	Open Space	1991	Fish & Amphibian

Table 2: Weather on day of sampling

Pond	Land-use	Weather
Laurel Employment Park	Commercial	Cloudy, Overcast, Sun Emerged
Golden Triangle	Commercial	Sunny
Laurel Lakes Court	Commercial	Sunny
Muirkirk Road	Residential	Cloudy, Overcast, Sun Emerged
Country Meadows	Residential	Sunny, slightly cool
Briar Oak	Residential	Cloudy, Somewhat cool
Fairland	Open space	Sunny, relatively hot
Bowie Fleming	Open space	Cloudy, Overcast

Table 3: Observed human use on day of sampling

Pond	Land-use	Human Use
Laurel Employment Park	Commercial	None, Very turbid
Golden Triangle	Commercial	None
Laurel Lakes Court	Commercial	None
Muirkirk Road	Residential	None
Country Meadows	Residential	Tire in pond
Briar Oak	Residential	Recreational fishing hole
Fairland	Open space	Walking path around pond
Bowie Fleming	Open space	Car in water, recreational fishing

Table 4: Pond depth and secchi depth on day of sampling

Pond	Land-use	Depth (m)	Secchi Depth (m)
Laurel Employment Park	Commercial	2.13	0.34
Golden Triangle	Commercial	0.84	0.34
Laurel Lakes Court	Commercial	1.52	0.69
Muirkirk Road	Residential	0.46	0.23
Country Meadows	Residential	0.46	0.23
Briar Oak	Residential	2.13	0.21
Fairland	Open space	>2.44	0.57
Bowie Fleming	Open space	>2.44	0.69

Table 5: pH and temperatures on day of sampling

Pond	Land-use	pH	Temp (°C)
Laurel Employment Park	Commercial	7.04	29.1
Golden Triangle	Commercial	7.15	28.5
Laurel Lakes Court	Commercial	7.34	28.2
Muirkirk Road	Residential	6.93	28.9
Country Meadows	Residential	6.85	22.3
Briar Oak	Residential	7.23	24.4
Fairland	Open space	7.1	24.4
Bowie Fleming	Open space	7.2	27.4

Graph 1: Dissolved oxygen values vs. Zooplankton Density

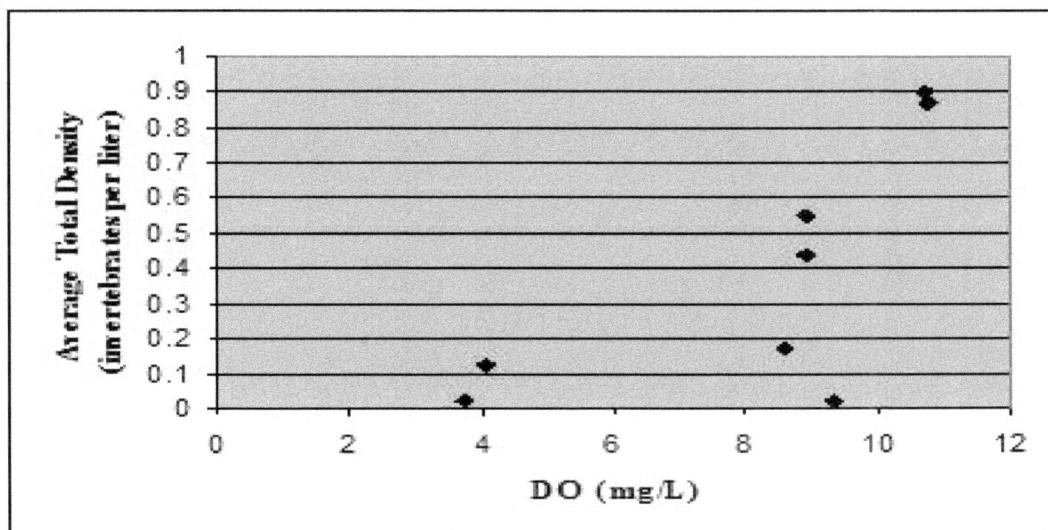

Chapter 5:

What do these findings mean?

Stormwater ponds, a somewhat new entity, are also a relatively new type of ecosystem. Since not much is currently known about these ponds, looking into the dynamics of these ponds is very important. Due to the similarities between lakes and stomwater ponds, there is an assumption that the dynamics of the two will also be similar.

Within lakes, cascades are important in structuring zooplankton communities. In general, larger fish prey on smaller fish, which prey on zooplankton. Larger zooplankton are also known to prey on their smaller counterparts. All zooplankton feed on algae. The removal or addition of any of the players in this cascade can change the dynamics. For example, the addition of fish will possibly lead to a decrease in zooplankton. This decrease in zooplankton may then lead to an increase in algae. This same cascade may take place in stormwater ponds, such as, those in this study.

Almost all the ponds had a definite presence of fish. This is a large factor. The remaining ponds with no recorded fish species, may very well have them, although not recorded due to an absence of data. A closer look at these ponds and the actual abundance of fish would be needed to determine their affect on the zooplankton communities in these ponds. This is due to the fact that a very large number of fish are usually needed to have an effect on zooplankton density[15].

The next step of the cascade is that of large zooplankton having an effect on smaller zooplankton[15]. Both copepods and cladocera are considered large zooplankton.

Rotifers are a smaller type of zooplankton[11]. *Chaoborus* would also be included in the large zooplankton category. When looking at zooplankton density across all sampled ponds, the total density of larger zooplankton (cladocera, copepods, and *Chaoborus*) compared with that of rotifers (smaller zooplankton) was greatest for all ponds except Muirkirk, a residential pond. However, when looking at microzooplankton vs. macrozooplankton, the microzooplankton outnumber macrozooplankton in all ponds except Golden Triangle, a commercial pond, and in Fairland, an open space pond, where their densities were equal. This is mainly due to the presence of cyclopoid larvae, which fall into the copepod category and are referred to as microzooplankton.

Continuing the cascade, there should be a relationship between zooplankton and algae. Chlorophyll analysis was done to estimate algal biomass for these ponds. It would be expected that those ponds with the greatest zooplankton density would have the lowest chlorophyll values. This however was not true. The largest chlorophyll value was actually coupled with the highest total zooplankton density. The larger zooplankton may be feeding on smaller zooplankton, which may in turn feed on the smallest type of zooplankton, protists (not true zooplankton). This being said, a smaller protists density would result in a larger than expected chlorophyll value.

The trophic cascade takes a top down approach to dynamics in water bodies. A competing theory to the trophic cascade is that of bottom up control. This may come into play if predation is not a major factor. The nutrients in the pond may limit the amount of algae in the pond, which could lead to a decrease of zooplankton. Due to the optimal chlorophyll value for rotifer density, seen in a couple of the ponds, these water bodies may be adhering to the bottom up theory and not that of the trophic cascade.

Larger chlorophyll values, signaling a larger algal biomass may also have its own effect on the zooplankton dynamics of a pond. Rotifer growth and reproduction increase with increased food availability[15]. Since rotifers feed on bacteria, as well as protists and algae, a greater algae presence would lead to a greater presence in rotifers. As mentioned before, the greatest chlorophyll value was coupled with the greatest overall density in Muirkirk Road pond. This chlorophyll value is also matched with a greater rotifer density. Rotifers in this particular pond seem to have been able to take full advantage of the algae present. The correlations between algae and rotifers in the remaining ponds are less straight-forward. Several of the rotifers identified in the study are known to be found in oxygen poor water in midsummer (*Asplanchna, Polyarthra, Filinia, and Keratella*)[11]. All samples were taken during mid to late summer. All samples were also taken in the open water area of the pond. Two types of zooplankton, *Asplanchna* and *Brachionus*, are known to only be found in the open water (limnetic) zone[11]. Both were found during the course of this study (*Brachionus* in two ponds and *Asplanchna* in five ponds) verifying this information for stormwater ponds in addition to lakes. A correlation was found between the density of rotifers and lower secchi depth values. This could be due to migration and predator avoidance. Zooplankton generally remain at deeper depths during the day to avoid being seen and eaten by fish. However, with a lower secchi depth, not much light is penetrating into the water, creating a dark environment, even near the surface of the water. The rotifers are then able to inhabit the surface waters without worrying about predation. This low secchi depth may be due to increased turbidity.

The largest populations of cladocera are found between temperatures of 6-12 °C. Due to the summer sampling, temperature of all the ponds was at least two times greater or more than this range. The lower cladocera densities may be due to this fact. Most species however are found within a pH range of 6.5 to 8.5[11]. All ponds fell within this range. A total of nine out of a possible fourteen types of cladocera, were represented in the study. Once again, all known open water genera were found (*Bosmina, Diaphanosoma, and Ceriodaphnia*)[11], again showing a similarity between lake characteristics and stormwater retention ponds. No correlations were found between the chemical variables and cladocera density, giving no new information regarding copepods and stormwater ponds.

Unlike their counterparts, copepods were found to respond to more of the typical variables which affect zooplankton abundance. The greatest copepod densities were found with the greatest oxygen availability as well as at the greatest phosphorus concentration. Conversely, the lowest copepod densities were found at the lowest phosphorus and oxygen levels. Correlations between copepod densities and the various chemical as well as physical characteristics recorded were stronger than that between other zooplankton group densities or total overall densities. From the newly discovered correlations, information about copepods in retention ponds can be inferred. They proliferate with riparian cover as well as algal presence, however, they appear to have an optimum level of both. Aside from oxygen and phosphorus, conductivity also correlated with copepod density. The less dissolved nutrients available, the better copepods can survive. This may be due to other pollutants in the water (such as metals) and well as low dissolved oxygen values.

Chaoborus, which do not fall into a formal zooplankton group, were only found in one pond. This pond also had the greatest pH and lowest conductivity levels. Chaoborus, like copepods, do better in water with fewer nutrients. They also seem to favor more basic waters. This is true for the stormwater ponds in the study, but may or may not be true for formal lakes.

Diversity was also found to have some correlation with a couple of chemical variables. Just like its effect on density, increased oxygen has a positive effect on diversity. Similarly, lower levels of conductivity parallel increased diversity. This signifies that oxygen as well as conductivity are important variables in the dynamics of stormwater ponds. Zooplankton, similarly to other members of the Kingdom

Animalia, need oxygen. They use it to fix nutrients to forms that they may be able to use. A greater conductivity can lead to less oxygen which is available for use.

All ponds were categorized based on land use. This is due to an assumption that ponds within a particular land use will receive similar runoff. The generated runoff may well contain toxins which can have an affect on the zooplankton community. The hypothesis set forth prior to completion of this study was as follows: numbers of tolerant zooplankton (as well as numbers of zooplankton in general) would be greatest in commercial ponds, followed by residential and then open space. In terms of diversity, open space ponds would have the greatest diversity followed by residential, and lastly commercial ponds. With regards to abundance of zooplankton (in general, not just tolerant genera), the density values were greatest in residential ponds, with open space ponds coming in last. When looking at diversity, the same was found to be true.

Residential ponds were found to be on top when looking at both abundance and diversity. This suggests that the makeup of residential runoff is able to both sustain and propagate zooplankton. The oxygen content of two of the three residential ponds was very high, in relation to the other sampled ponds, possibly giving rise to the large numbers and diversity. Phosphorus and ammonia content among all eight ponds was varied, with commercial ponds having the greatest average value. These ponds were found to be moderate in terms of phosphorus and ammonia values. Neither phosphorus nor ammonia were found to be a major factor in abundance or diversity for overall zooplankton. Phosphorus however was found to be related to copepod density.

Bosmina, a tolerant zooplankton, was discovered in only half (four) of the sampled ponds. It was found however to be present in all land uses. In the Laurel Employment Park pond, a commercial pond, four *Bosmina* were found within all samples. Two ponds belonging to the residential land use category, Briar Oak and Muirkirk, were found to have *Bosmina* present. Lastly, Bowie Fleming, an open space pond had fourteen *Bosmina* found in total. Overall, the greatest number of *Bosmina* were found in residential ponds, open space next, and then commercial ponds. Another tolerant zooplankton, *Asplanchna priodonta*, was found in Laurel Lakes Court, a commercial pond, but only one individual was found. Not only do residential ponds rank first in total abundance and diversity, but also in abundance of tolerant zooplankton.

Beta diversity, the difference between two ponds or two land uses, depending on what is being compared, was calculated. The differences found between land uses, in terms of types of zooplankton not found in the other land use, was similar to those values found when comparing ponds within the same land use category. This shows that diversity is more than likely not related to land use but varies on an individual level, from one pond to another. However, multivariate analysis showed that several ponds were similar to each other based on both environmental data alone, as well as the presence or absence of zooplankton. Open space ponds were the most similar in both cases (although two commercial ponds also clustered based on both environmental data alone and the presence or absence of genera). This suggests that receiving almost no commercialized runoff leads these ponds to be more similar. Both commercial and residential ponds do receive runoff. This runoff however, likely varies from pond to pond, causing stormwater pond communities to vary from pond to pond.

Since residential ponds were found to have the most abundance and diversity, it seems as if the intermediate disturbance hypothesis may come into play. In 1978, Connell projected this hypothesis, which states that diversity reaches its maximum level under intermediate levels of disturbance[1]. These disturbances can be due to human use, nutrients, as well as floods or droughts. Too much disturbance may lead to a decrease in diversity (ex: commercial ponds), while an optimal level of disturbance (ex: residential ponds), causes zooplankton communities to flourish. Open space ponds, which generally receive a low amount of disturbance, may be lacking nutrients, which led to less zooplankton diversity.

When compared to the previous study of stormwater ponds by Karouna-Renier and Sparling in 2001, some of the findings were similar[8]. Although this study looked at metal loads in macroinvertebrates, findings of the water chemistry can be compared to this study. Water chemistry did not show a significant difference between land uses. This was also typically true in this study. Also, increased urbanization was found to increase turbidity. This increased turbidity could be the cause of lower secchi depth values, however, the lowest values in this study were found in two of the residential ponds. This does not adhere to the previous findings due to the fact that commercial ponds would be those with the most urbanization.

Chapter 6:

What is the take home message?

The dynamics of stormwater ponds are similar to that of natural wetlands. Although the trophic cascade was not adhered to, several other unstudied and unmeasured characteristics may be responsible for observed differences (or possibly bottom up control). A reoccurring theme stressed the need for oxygen. Higher dissolved oxygen content is beneficial for zooplankton in terms of diversity and abundance. Conductivity is also important to some types of zooplankton. However, copepods in stormwater ponds adhere to phosphorus theories relating to zooplankton.

In terms of land use, residential ponds had both the highest diversity and abundance. The runoff which they receive helps the community instead of hindering it. Values for these ponds were higher than the "natural" or open space ponds. Biodiversity in stormwater ponds was slightly affected by land use, but more by individual pond characteristics (ex: oxygen and conductivity).

About the Author

MICHANNA TALLEY is a lawyer in Greenville, South Carolina. She is a graduate of Stetson University College of Law located in Tampa Bay, FL. Michanna also holds both a Master's and a Bachelor's degree in Biology from Howard University in Washington, DC. She also holds Graduate Certification in the field of Public Health, specifically in Epidemiology and Biostatistics from Drexel University located in Philadelphia, Pennslyvania. Prior to entering law school, Michanna was an Adjunct Instructor teaching General Biology 101 and 102 and prior to that was a Nucleic Acids Microbiologist working in the field of Biodefense in the DC metropolitan area. Even with her legal career, Michanna has continued to keep one foot in the scientific field by teaching online for University of Phoenix as well as local universities. Michanna also is able to have an outlet for her creative side with her business, Jazi Gifts (www.jazigifts.com).

References

1. Connell, J.H. 1978. Diversity in tropical rainforests and coral reefs. Science 199:1302-1310.

2. Dodds, W.K. 2002. <u>Freshwater Ecology: Concepts and Environmental Applications</u>. Academic Press, San Diego.

3. EPA. 1999. Storm Water Technology Fact Sheet: Wet Detention Ponds. United States Environmental Protection Agency: Office of Water 832-F-99-048.

4. Ferguson, B.K. 1998. <u>Introduction to Stormwater</u>. John Wiley & Sons, Inc., New York.

5. Hairston, N.G. Jr., and Hairston, N.G. Sr. 1997. Does food web complexity eliminate trophic-level dynamics? American Naturalist 149: 1001-1007.

6. *An Image-Based Key To The Zooplankton Of The Northeast (USA)*. Version 2.0. CD-ROM.

7. University of New Hampshire Center for Freshwater Biology (http://cfb.unh.edu). 2005.

8. Karouna-Renier, N.K., and Sparling, D.W. 2001. Relationships between ambient geochemistry, watershed land-use and trace metal concentrations in aquatic invertebrates living in stormwater treatment ponds. Environmental Pollution 112: 183-192.

9. Oertli, B., Joye, D.A., Castella, E., Juge, R., Cambin, D., and Lachavanne, J. 2002. Does size matter? The relationship between pond area and biodiversity. Biological Conservation 104: 59-70.

10. Pedros-Alio, C., and Brock, T.D. 1985. Zooplankton feeding dynamics in Lake Mendota: short-term versus long-term changes. Freshwater Biology 15: 89-94.

11. Pennak, R.W. 1989. <u>Fresh-Water Invertebrates of the United States</u>. John Wiley & Sons, Inc., New York, NY.

12. Richmond, Mark. "Re: Stormwater Pond Info." E-mail to Michanna Talley. 9 September 2005.

13. Simon, Judith. "Re: Stormwater Ponds Fish Data." E-mail to Michanna Talley. 14 September 2005.

14. Stahre, P., and Urbonas, B. 1990. <u>Storm-water Detention: For Drainage, Water Quality, and CSO Management</u>. Prentice Hall, New Jersey.

15. Wetzel, R.G. 2001. <u>Limnology: Lake and River Ecosystems</u>, 3rd edition. Academic Press, San Diego.

16. Whittaker, R.H. 1972. Evolution and measurement of species diversity. Taxon 21: 213-251

17. Williams, P., Whitfield, M., Biggs, J., Bray, S., Fox, G., Nicolet, P., and Sear, D. 2003. Comparative biodiversity of rivers, streams, ditches, and ponds in an agricultural landscape in Southern England. Biological Conservation 115: 329-341.

18. Wood, P.J., Greenwood, M.T., and Agnew, M.D. 2003. Pond biodiversity and habitat loss in the UK. Area 35.2: 206-216.

19. Wood, P.J., Greenwood, M.T., Barker, S.A., and Gunn, J. 2001. The effects of amenity management for angling on the conservation value of aquatic invertebrate communities in old industrial ponds. Biological Conservation 102: 17-29.

20. Yakovlev, V. 2001. Zooplankton of subarctic Imandra Lake following water quality improvements, Kola Peninsula, Russia. Chemosphere 42: 85-92.